The Permaculture Student
1

Written by Matt Powers

The Permaculture Student

Much of this textbook and workbook is inspired by the work of Geoff Lawton, Geoff's online permaculture design course and the work of his predecessors: Bill Mollison, David Holmgren, Masanobu Fukuoka, & PA Yeomans. It has been adapted to a general audience. This book works easily as a supplement to any U.S. middle school science class or its international equivalent. This book can also be seen as a crash course in permaculture, suitable for any adult.

This series was created to bring ethical design thinking to the education of children via action-oriented, positive, hands-on activities that connect a broad range of sciences: Agriculture, Horticulture, Ecology, Chemistry, Architecture, Landscape Design, Nutrition, and Biology

Please Note: Just as food that nourishes one person causes an allergic reaction in another, the same concept of situational complexity applies to soils, dams, medicine, mushrooms, and more. In Permaculture, complexity is embraced with the understanding that every situation and biome is unique. The information in this book represents research from sources listed - it is an educational and informational resource and does not represent any agreement, guarantee, or promise by any party associated with the creation or editing of this book. The publisher, editors, and author are not responsible for any negative or unintended consequences from applying or misapplying any of the information in this book.

All Inquiries:
PowersPermaculture123@gmail.com
28419 SE 67th St
Issaquah, WA 98027

Illustrations
Page 8, "Light Bulb" by Matt Powers.
Pages 2-3, 5, 26, 34, 35, 41, 48 bottom, 51, 62, 64-65, 68, 79, 81 by Wayne Fleming.
Page 89 by Lyric Piccolotti.
All other art by Brandon Carpenter.

Formatted by Thomas Mitchell of Byblos Media.
Published 2015, PowersPermaculture123.
Copyright © 2016 by Matt Powers.
2nd Edition 2016.

Table of Contents

Chapter I 5
Introduction

Chapter II 15
Nature

- **Behaviors of Nature** 16
- **Elements of Nature** 21
- **Soil** 23
- **Fungi** 27
- **Trees** 29
- **Climate** 32

Chapter III 35
Permaculture Design

- **Observation** 36
- **Planning** 47
- **Action:**
  - Soil 52
  - Plants 61
  - Animals 68
  - Aquaculture 70
  - Earthworks 72
  - The Home 76

Chapter IV 79
Permaculture and the Future

Index 82

Chapter 1

Introduction

What is permaculture?

Permaculture began simply as permanent agriculture, an **ethical**, **regenerative** food system, but it was expanded to be an entire ethical design system to develop and maintain permanent cultures. It focuses on using regenerative **energies** in the ways that nature already does but by design, capturing and utilizing all, including **potential energy**. Permaculture works with, benefits, and extends the patterns of nature.

Providing food regeneratively is not new. Many cultures have had regenerative **elements** and understanding. Every person today has had ancestors that have lived harmoniously enough with nature to survive. These ancestors had no access to the scientific research, modern technology, and the plant and animal **diversity** we have today. These ancestors were keen observers of the patterns of nature and restricted their own consumption of resources to keep those resources renewable. With our current understanding of both past and present permaculture design, we can create **resilient** & sustainable systems to enjoyably provide for all our needs locally and globally in a way that benefits nature.

> **Ethical:** ideas and actions that do no harm to people or the environment
> **Regenerative:** restoring, regenerating, and improving continuously
> **Energies:** forces that can be used to power a process i.e. sun, heat, wind.
> **Potential Energy:** resources or elements that have the potential to be used to create energy: water, gravity, firewood, etc.
> **Element:** a part of something greater than itself i.e. a tree in a forest.
> **Diversity:** the amount of variety
> **Resilient:** able to resist or recover quickly from stress or damage

Design Ethics

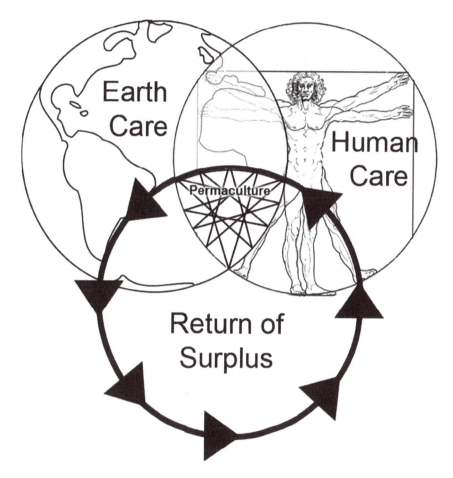

Earth Care
Care of all living and nonliving things on earth

•

Human Care
Care of all humanity with self-reliance
and community responsibility

•

Return of Surplus to the other Ethics
To facilitate the care of the first two ethics
- commerce, trade, charity, sustainability and wilderness -
Also known as: Care of the Future

•

All designs have to find an ethical balance, the overlap between
the three ethics. When balanced, designs are always beneficial
to the earth and all life.

The Prime Directive

"The only ethical decision is to take responsibility
for our own existence and that of our children. MAKE IT NOW."
- Bill Mollison, *Permaculture: A Designer's Manual*

The Problem Is The Solution

Permaculture sees problems as opportunities for improvement. For example, through design, unwanted waste can become a valuable resource. The potential value of the resource all depends on the size of the problem. Surplus wind can become energy with a turbine. Surplus water can become productive ponds or hydroelectric power. Surplus sun can become solar power. We are only limited by our imaginations.

Permaculture in Landscape and Society

1. Preserve and Protect Remaining Untouched Wilderness
2. **Rehabilitate Degraded** Land
3. Create Our Own Complex Living Environments
   (Mollison, <u>Permaculture: A Designer's Manual</u>, 1988).

Rehabilitate: to bring something back to the way it was before
Degraded: damaged, lowered in quality and function

Work With Nature

*observation journals

Recognizing the way nature works is the first step to working with nature. Using nature's methods, we can use less energy to accomplish our objectives and benefit the earth as well. When we allow in beneficial bugs, fungi, & "weeds", we aren't just saying no to pesticides, fungicides, & herbicides; we are agreeing to work with nature's systems. All those elements are critical elements in a thriving ecosystem and are necessary to make healthy soil, food, and people.

Everything Gardens

In balanced ecosystems the inputs and outputs of each element enhance its environment. The mole and worm **aerate** the soil. The birds and browsers spread the edge of the forest by spreading seeds and fertilize their own food sources. "Weeds" like vetch and clover repair the soil as do almost all **legumes**. Weeds indicate what is missing from the soil life because every weed type brings in specific nutrients or minerals that the ecosystem is needing. **Degraded** landscapes are always trying to reestablish themselves. All life is pushing towards a greater expression of life. If we utilize nature's gardeners, we can make a fast, powerful, and positive change in the environment.

> **Aerate:** to add air to something
> **Legumes:** plants in the bean and pea family that often fix nitrogen. They are critical to all permaculture food forests and gardens.
> **Degraded:** to drive evolution backward, lowered in quality and function

10 | THE PERMACULTURE STUDENT

Make the least change for the maximum effect

The best designs are a favorable balance of input to output. A good design should use the the least amount of energy or resources possible for the most benefit. For example, in order to prevent a **frost pocket**, removing lower tree branches instead of the entire tree allows cool air to drain rather than be trapped. In Australia, if you insulate your ceiling you will get a 40% reduction in heating and cooling costs.

A simple log dam can partially re-route a creek or stream to irrigate a garden bed to great effect.

Frost Pocket: an area where cold, still air collects, usually in a shady hollow.

Conventional? Organic? Permafood?

There is a lot of confusion over what "organic" means. Most people think it means no **chemical sprays** and no **GMOs**. While those are the main ideas behind the regulations, it is more complex than that; it has different sets of rules for different types of agriculture. In the U.S., the term 'organic' is an **FDA certification** that governs the way food can be grown on a certified farm though it is not an indication of how healthy the food is. Also note that anyone can grow organically, without artificial chemicals, and doesn't need certification to do so.

Some commercial agriculture have no specific organic **standards** to guide them. In permaculture systems, nutrition is encouraged. Permafood, or food from a permacul-ture system, becomes healthier each year as the soils are enhanced. Industrial foods use synthetic chemicals and unnatural or unethical processes; permaculture systems imitate nature and can prove their nutritional superiority.

Foods can be liquified or squeezed for juice and tested with a **refractometer** for starch (or sugar) levels. Starch levels indicate how well a plant is photosynthesizing, how effectively it is exchanging nutrients with the **rhizosphere,** and how dense its nutritional profile is in a general sense.

Beyond nutrition, the taste of foods raised in permaculture-inspired systems is prized by chefs and cherished by home growers.

> **Chemical Sprays:** chemical-based fertilizers, fungicides, herbicides, and pesticides.
> **GMOs or Genetically Modified Organisms:** an organism that has had its genes altered in a laboratory, usually crossing genes of a different species.
> **FDA:** Food and Drug Administration, a regulatory branch of the U.S. federal government
> **Certification:** confirmation or recognition by the regulatory body of certain qualities or characteristics
> **Standards:** rules or code of conduct
> **Refractometer:** a device that tests liquids by light refraction for their starch/sugar levels. Used commonly by commercial honey, wine, and juice operations.
> **Rhizosphere:** the area below the surface of the soil where plant roots grow

Top World Problems

Water Scarcity: Increasingly drought is threatening worldwide food production while governments, corporations, and individuals pump water out of aquifers at a rate that will never be recharged in our lifetimes or our children's lifetimes. Industrial need for water has also increased exponentially as scarcity has further stressed natural sources of clean, fresh water putting the needs of these industries before the needs of nature and future generations. Fresh unpolluted drinking water sources are rare. We need to see fresh potable drinking water as the world's most precious mineral.

Soil Degradation: Soil is the source of all life in the environments that humans and most life exists. Even the fertile areas of the ocean have their own kinds of soil. Topsoils worldwide are eroding faster each year; half our topsoil has been lost in the past 150 years. Farming practices, lack of soil science understanding, global market demands, and climate change have all contributed to the loss of topsoil, but solving the issue is more important than its causes. Permaculture techniques build soil by imitating the processes of nature.

Deforestation: As the forests are removed, the topsoils are washed or blown away. Habitat for organisms is also lost, causing extinction of species and sometimes the loss of the eco-systems themselves. Forests have always provided the clean water, air, food, and animals that support human populations. Without forests doing this work **passively**, human civilization have had to turn to increasingly expensive means of doing these jobs themselves. If everything needed was produced locally and regeneratively, no more forests would need to be clear cut. Permaculture design helps us build forests that can last centuries and even millennia by observing and obeying the patterns and systems of nature.

Pollution: Pollution is a very serious, growing problem though it can be addressed. Almost all the junk we release when we burn fossil fuels can be returned to an **inert** state with composting or fungi. Even radioactive waste can be eaten by fungi. Humanity's waste is a huge problem whether it is noise, air, soil, or water pollution, but pollution itself is a design problem, a mismanagement of resources, and the **excess** waste can be taken back into a natural cycle as long as we do not create and release substances that do not cycle (like **DDT** for instance). We have to refuse to use or boycott dangerous chemicals like Agent Orange and DDT. All waste must be recyclable in a permaculture design, and every site must be responsible for their own waste.

> **Passively:** working without active stimulation
> **Inert:** non-reactive, harmless
> **Excess:** more than a system can use
> **DDT:** Dichlorodiphenyltrichloroethane; a banned pesticide.

Chapter II

Nature

Behaviors of Nature

Diversity

Diversity is the amount of variety in a system; biodiversity is the variety of life in an ecosystem. The greater the interaction in a system, the more resilient and stable it is. The more interaction between elements of an ecosystem, the more stable it is. Stable systems are predictable and **accumulate** resources which increases **fertility**. Increased fertility leads to more rich and diverse ecosystems over time until climax is reached.

"Ecosystems are complex systems; they are finely integrated and inter-related and inter-dependent in ways that we barely understand. They have a structure, and they function to maintain life in all its aspect in good health e.g. clean water, build soils, maintain plant and animal fertility, maintain air quality and climate stability."
— Rosemary Morrow

Life systems are based on diversity and perpetuate through diversity.

> **Accumulate:** to gather an increasing quantity of something
> **Fertility:** the potential for life

When a legume tree starts to grow in an area of low fertility and diversity, it starts a chain reaction. Its leaf litter covers the soil and nourishes it as it decomposes. Its seeds start to spread the growth of more trees and to create shade. Nearly all legumes also fix atmospheric nitrogen into the soil using their root nodules to interact with soil bacteria. The seeds and leaves of the plant return the most nitrogen to the soil, increasing fertility. As the leaf litter creates fertile soil and the trees create shade, water is retained, animals are fed, soil decomposers fed, the soil fertility increases, and the life diversifies.

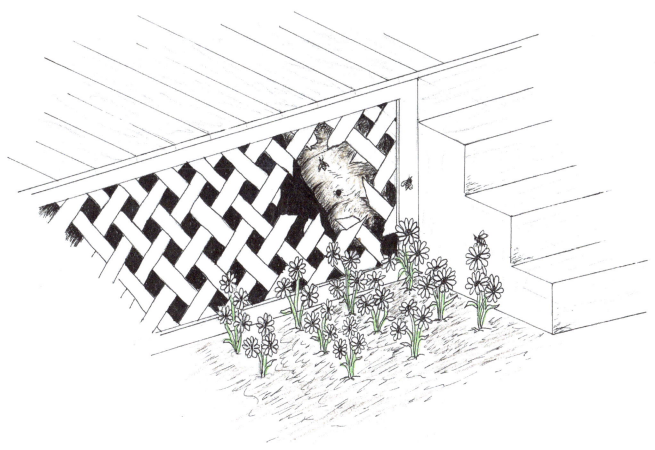

Niche

Niches are roles or openings in the ecosystem's diversity. It can be any life form performing any functional role in an ecosystem. Permaculture fills niches by design with desirable life forms. A break in a front porch lattice might be a problem for the home-owner, but for bees, it might be a perfect location to be close to the garden.

Cycles: Niches in time

Cycles are patterns that work in stages over time. Each stage builds towards the next. There is no starting or end point; it is continuous. Cycle size is not a determining factor as they can occur microscopically inside a cell or within the global atmosphere. It is our job as designers to recognize, support, and/or moderate natural cycles in our system and world. Nature's cycles avoid waste accumulation. The waste of one stage becomes the resource for the next: grass eaten by cows passed on as manure is spread and picked over by chickens and then soil life to return as grass again. At each stage the waste is what is used by the next stage.

Snows pile up in winter and melt in spring to make streams that feed the new growth each year. This cycle powers the cold temperate climate system. Deciduous trees drop their leaves creating a thick blanket of **mulch** that protects roots and seeds from winter freezes. The fallen mulch composts over winter and becomes the topsoil for spring's growth.

> **Mulch:** organic material that is breaking down into rich soil: leaves, compost, sticks, bark. It is ideal to cover topsoil with to protect soil life.

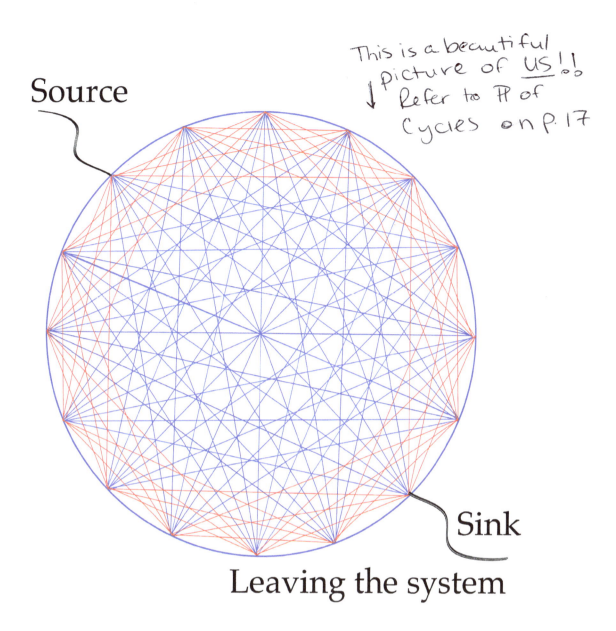

Source

Sink

Leaving the system

This is a beautiful picture of US!! Refer to P of cycles on p. 17

The Web of Life

In a healthy ecosystem, energy, water, and fertility enter through a source and cycle through as many mediums in the environment as possible before leaving the system in a sink. This can include animals, plants, soils, and even the atmosphere. Everything interacts, or shares interdependence, and cycles nutrients and energy. A well designed permaculture site traps and cycles energy and fertility indefinitely.

Forests can last thousands of years.

The Global Water Cycle

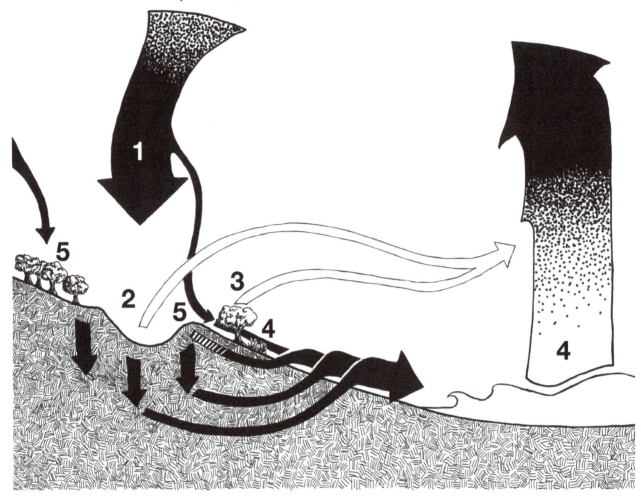

1) **Precipitation**: any form of water that falls to the ground from the atmosphere
2) **Evaporation**: the cooling process of water changing from liquid to gas. This is especially a problem with evaporation from the soil.
3) **Transpiration**: the physiological process of water moving through a plant and evaporating through its leaves, stems, and branches
4) **Evapotranspiration**: movement of the air or water that causes water to be lost when transpired water becomes a gas and becomes clouds.
5) **Condensation**: the warming process by which water vapor collects on any surface as it changes from gas to liquid. It is a way we can harvest water.

There are also cycles within cycles. A drop of rain is absorbed by the soil and tak-en up by a plant. That plant's leaf is eaten by an animal that later urinates on a distant patch of soil which feeds another plant, which is eaten by aphids, those aphids are eaten by birds, those birds defecate beneath a nightshade plant, a plant takes up the nutrient and moisture in its roots through the soil, and so on. Finally it leaves the system through evaporation or joining a larger water source. Water travels downward continuously towards sea level - however, when water reaches flat land it is pacified and soaks back into the soil.

Elements of Nature

The Sun

The Sun is the source of all energy. It powers our planet and all nature's processes directly or indirectly. The sun powers the engine of the planet from the core through the atmosphere. We rotate around the sun, and we are spin on an axis. The sun influences how everything grows and behaves on the earth and provides the context for processes that don't use light to exist.

Water | All Life

Living systems need water to survive. Starting any permaculture site requires analyzing how much water is on the site, how much precipitation there will be, and what times of year the water is readily available.

Aquaculture

Life-rich water systems provide an abundance of food perpetually, more than any land-based system.

Energy Source

Water is a potential energy source and has always been a power source for both natural and human-made systems. Storing water as high as possible on the land ensures the most potential energy. It is potential energy because passive bodies of water do not provide energy for human use without human intervention or a natural process.

It is our responsibility to recharge the aquifers we have drained, to restore the watersheds we have damaged or removed, and to clean our rivers and streams of the toxins in them.

Wind

Wind is an amazing phenomenon. Though it is invisible largely, it carries silt, seeds, nutrients, bugs, and birds from far distances. It also prevents fungal diseases, cools, causes trunks to thicken, and even prunes trees. Wind can be turned into electricity, lifted above an area with a forest belt, slowed down by a windbreak, or channeled with a wind tunnel. It can be a destructive force if the site is not designed properly.

Soil

Soil is the largest, most diverse, and complex life system known to science. It is less understood than space. Only recently has science been able to tackle the science of soil.

Healthy soil makes for healthy plants. Healthy plants provide clean water, clean air, and abundant food. Organic carbon is the foundation for all the structure of the life in our ecosystems, but plants and animals also need a proper balance of nutrients available in the soil. Having a large diversity of organic matter and soil life provides all the nutrients needed.

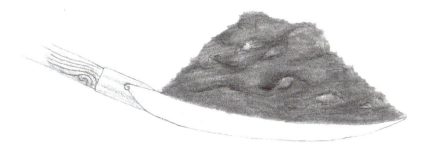

> **Primary Nutrients:** Nitrogen (N), Phosphorus (P), Potassium (K)
> **Secondary Nutrients:** Calcium (Ca), Magnesium (Mg), Sulfur (S)
> **MicroNutrients:** Boron (B), Copper (Cu), Iron (Fe), Chloride (Cl) Manganese (Mn), Molybdenum (Mo), Zinc (Zn)

Soils host millions of organisms like bacteria, fungi, nematodes, and protozoa, many of which are not identified yet. There is also air and water in the soil which most organisms need to survive. These small organisms can be seen with a microscope and studied. Their activities retain water and provide nutrients to the plants and each other.

Plants have a preferred ratio of Fungi to Bacteria (F:B). Annuals, vegetables, and grasses prefer bacterial dominated soils. Perennials, trees, and shrubs prefer fungal dominated soils. All old growth forests are in acidic, fungal dominated soils.

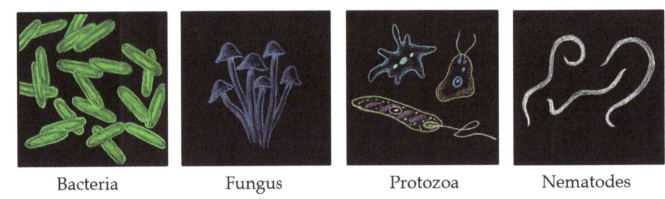

Bacteria   Fungus   Protozoa   Nematodes

Returning organic matter to the soil is the only way to maintain the fertility of the soil. **Biocides**, **fungicides**, **herbicides** and **pesticides**, destroy the healthy soil life that creates healthy foods. Fertilizers that are only primary nutrients lack vital organic carbon, sec-ondary nutrients, and micronutrients. Building soil the way nature does provides all that is needed for abundant healthy foods.

In the undisturbed natural world, soil is created through a combination of processes: **weathering**, chemical break down and decomposition. The physical action of a glacier over bedrock, water over stone or wind through a canyon are all examples of weathering. Fungi break down rock and complex forms of organic matter while bacteria break down simple, non-complex organic matter like simple sugars. Fungi acidify the soil while bacteria make it more alkaline. The four basic components of soil are clay, sand, silt, organic matter and organisms. Soil is biological, fungal, mineral, and bacterial.

> **Biocides:** a substance that kills living things, usually made with synthetic chemicals that usually persist for years in the environment.
> **Herbicides** target plants. **Fungicides** kill fungi. **Pesticides** kill a widerange of life.
> **Weathering:** natural forces and physical processes that break down rock and other elements into soil.

The Soil Food Web

The soil food web is a map of the interconnections and cycles of soil life. Balance in the soil food web occurs when both fungi and bacteria are diverse and when organic matter is freely available because all the other levels of organisms need those elements to thrive. When all levels of the soil food web are active, they make available non-soluble sources of nutrients for plants to take up, create soil structure, and retain moisture and nutrients. Soil life is the key to fertility in the soil.

"It can't be soil without life." - Dr. Elaine Ingham

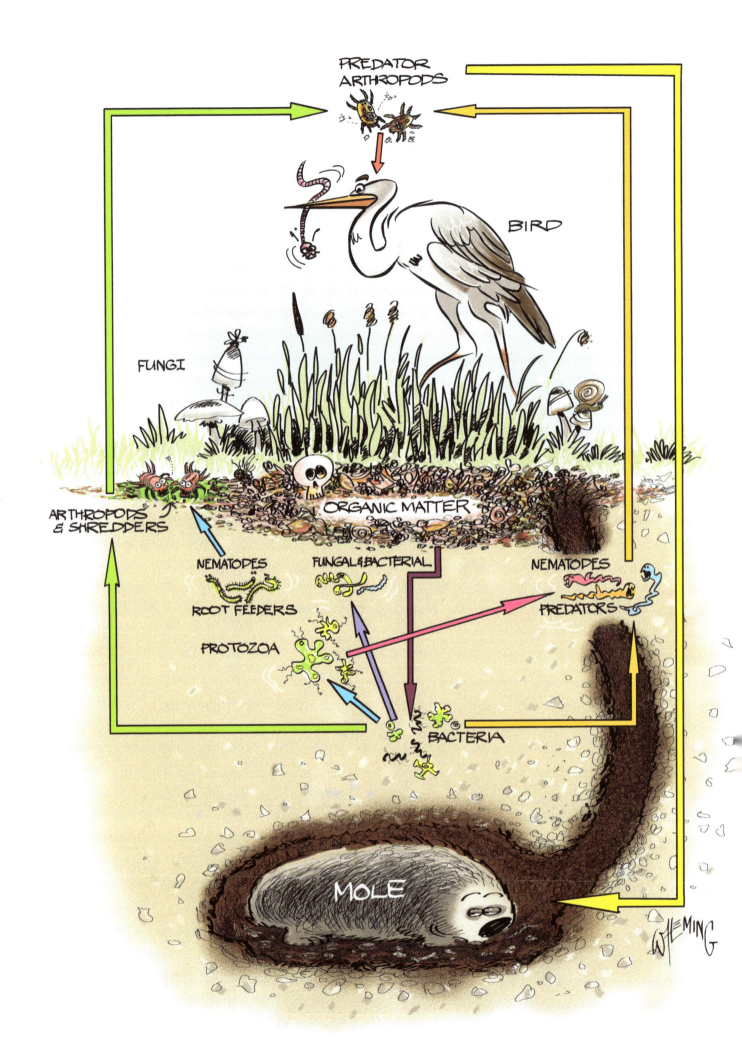

Fungi

Fungi are an essential component of all ecosystems, most natural cycles, all life, and all soils. They decompose organic matter and minerals in aerobic conditions, spread disease in oxygen depleted conditions and form a communication and nutrient exchange network in the soil that can go for miles. Fungi break down the lignin fibers of wood; almost all trees prefer to partner with fungi. Mycorrhizal fungi's hyphae work with plant roots to trade nutrients from soil microorganisms for plant **exudates**. Those exudates are consumed by the fungi and bacteria. Nematodes and protozoa feed on the fungi and bacteria, and they excrete plant readily-available nutrients into the soil as waste. By putting out exudates, plants attract fungi and bacteria to attract nematodes and protozoa that will feed the plants with their wastes. The plants put out the exact exudates to attract the exact **nematodes** and **protozoa** they need to generate the exact food they need to thrive. Without a fungal hyphae protective network in place around the plant, they would not be safe from root feeding nematodes and predatory soil life. The plants also would not be able exchange their exudates to attract the correct fungi and bacteria.

Fungal dominated soils are essential for longterm, regenerative growth; all the old growth forests are all growing in fungal dominated soils.

> **Hyphae:** an element of fungi: the long, thread-like branches
> **Exudates:** mostly carbohydrates (starches and sugars) and some proteins
> **Nematode:** a worm-like, multicellular microscopic animal that feeds on fungi and bacteria
> **Protozoa:** a unicellular microscopic organism that feeds on fungi and bacteria

"Fungi are the interface organisms between life and death." - Paul Stamets

Mushrooms are the fruit of the fungus. Many mushrooms are poisonous to eat, but there are also many that are delicious and nutritious. Sometimes it can be difficult to know which mushrooms are safe; many look very similar. Eating wild mushrooms can be very dangerous, even if they look like a store-bought mushroom. Learning with an experienced mushroom forager is critical.

Fungi break down trees. When this happens, the wood is often called "punky". Forests grow on fallen forest. Without fungi, there would be no forest soils.

Mycorrhizal fungi form **mycelium**, a communication network in the soil with their hyphae. These networks can reach for miles. The hyphae are the exchange pathways for nutrients and minerals. Trees struck by a pest will communicate through the mycelial network, and trees miles away will begin to adapt a **resistance** to that pest.

We are all connected!

> **Mycelium:** the body of the fungus
> **Resistance:** the ability to resist influence from an outside source

Trees

Humans have always relied upon trees. They provide food, clean air and water, shade, building materials, mulch, **habitat**, historical records, windbreak, fiber, medicine, and more. Without trees we wouldn't have the diversity of plants, animals, materials, and resources necessary to support human life. We are in a **symbiotic** relationship with trees. Trees interact with every level of an ecosystem.

> **Habitat:** living environment for an organism
> **Symbiotic:** interdependent

Trees and Wind

Trees cool hot winds, warm cold winds, and slow winds down which causes them to drop the nutrients and particles they were carrying. When winds pass over trees, they spiral, and a protected area just after the trees is formed.

Trees and Water

Through transpiration, trees release water back into the atmosphere. Through condensation, trees capture water from the atmosphere. Trees absorb water through their roots as well. Forests on mountaintops hold moisture in the air and soil. Their interactions with the atmosphere cause precipitation. If the mountaintop forests are cut, precipitation, cloud cover, and habitat will all disappear. Deforestation causes desertification.

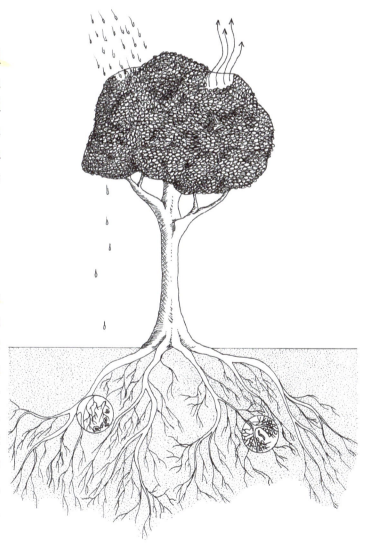

The layers of a forest must all be filled, or nature will fill any empty niche for us, usually with what is called a "weed". The layers are: the climax or canopy of large trees, the understory or trees in the shade of climax trees, shrubs and bushes, herbaceous, low-er herbaceous (in cold temperate), ground cover of creeping plants, vertical or climbing vines, clumpers or plants that grow by division like bamboo, the fungal layer, and the root layer or rhizophere - the area around the roots. In the tropics there can be up to two layers of palms as well.

If you understand the way a forest grows, you can design your own.

Weeds

Weeds are repair mechanisms. They show up to repair the land. They also will fill any gap left in the layers of the forest. Deep tap rooted plants show up in compacted soils. Hair net rooted plants show up in loose soil. Fire weeds show up after a fire to return phosphorus to the topsoil. Instead of pulling weeds as many do, it is better to chop and drop them in place before they set seed, so the nutrients they are accumulating for the soil can be added to the topsoil as mulch. The decomposing support plant will both build new soil and repair the deficient soil with their new layer of mulch. This speeds up the natural cycles that build soil.

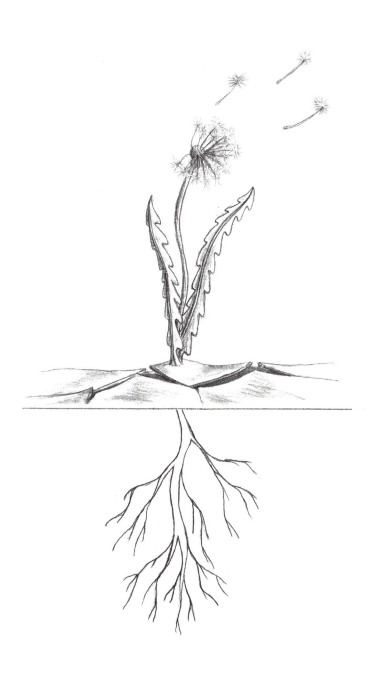

Climate

Broad Climate Zones

Though each ecosystem on earth is unique, there are general similarities that allow us to categorize and study them as broad climate zones.

> **Temperate:** extends from the polar zone to the mediterranean, warm to cool to cold.
> **Tropics:** hot and humid equatorial zone between the Tropic of Cancer and the Tropic of Capricorn
> **Dryland:** high evaporation zone found in all areas
> **Polar:** extending out from the poles, extremely cold and dry tundra with no warm summers or trees.

Major Landscape Profiles

Humid: high moisture content, rounded hills and mountains
Arid: low moisture content, angular landscape, strong winds, high evaporation, high mineral and nutrient content in the air and ground

Minor Landscape Profiles

Volcanoes: alkaline soil, steep slopes, fertile ring plain
High Island: half humid, half dry, rain shadow effect
Low Island: fresh water lens under the surface, strong winds
Wetlands: high water table, difficult for vegetable production
Flatlands: strong winds, no potential for gravity-powered watering
Estuaries: tidal flows, marine aquaculture, abundant in nutrient
Coasts: alkaline, salt winds, fast draining soils, lack of soil nutrient

Microclimates

Microclimates are formed when an area gets more or less energy than the surrounding area. More sun may mean drier, warmer conditions longer into the cold season. More water can mean more fertility in the dry season. Windbreak provides shelter for tender plants. Microclimates raise diversity and stretch an area's possibilities. They can be created with almost anything and can be found almost anywhere.

Microclimates in cold climates often trap heat to protect sensitive crops. In the example above, a pond is used to reflect the sun, a boulder half-buried in the ground is used for thermal mass behind the tree, a windbreak of trees and plants block cooling winds, and a bed of native, hardy plant skirts our valuable fruit tree.

Chapter III

Permaculture Design

Observation

Observation is perhaps the most powerful tool we have to unlocking the power of natural systems. Though our own observational skills may be limited, we can use techniques and tools to enhance our skills. Over time and with experience, reading the landscape becomes easier.

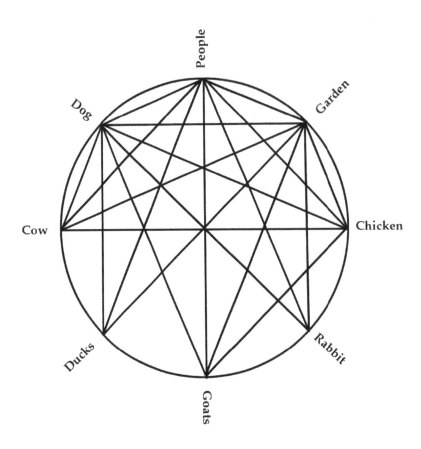

Every Element Has Multiple Functions and Supports

In nature every animal, plant, microorganism, and process has multiple functions and supports. The more connections (inputs and outputs) between elements in a system, the more sustainable a system is.

For example, chickens eat a wide variety of foods, so they can survive in a wide variety of climates and conditions. Chickens provide eggs, meat, feathers, bones, pest control, manure, scratching that is like light-tilling, chicks, and more. Our imaginations and observation skills are our only limitations.

Gravity

Gravity is a constant force that influences everything. It also has immense potential energy if properly designed. Recognizing how gravity influences a site opens possibilities of extending those current patterns or redirecting them. Using gravity as a force for power in a design can create an abundance of electricity, water storage, **aquaculture,** and almost anything else you can create using those **products**.

> **Aquaculture:** aquatic plant and animal farming
> **Products:** the result of a process

Altitude Effect

How high up we are in our atmosphere effects the climate. It's as if we travel away from the equator towards colder, temperate climates. This effect is important to remember when looking at any site at a high altitude.

"For every 100 meters [328~ ft] there is the effect of moving 1 degree latitude away from the equator" -Geoff Lawton

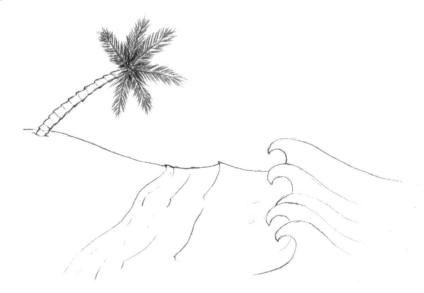

Maritime Effect

Bodies of water have ability to moderate the climate around them. Large bodies of water have the same effect but greater. The Maritime Effect causes mild winters and summers. It is often fantastic for growing certain foods (if safe from winds carrying salt).

Continental Effect

The inverse of the Maritime effect is also true: the further away from the large bodies of water on earth, the hotter the summer and colder the winter.

Rain Shadow

As rain clouds approach a mountain they drop their rain on the side they approach from, and over time one side of the mountain is wetter and one side is drier. This is most easily observed with coastal mountain ranges. Looking at a piece of land and knowing where the prevailing winds and storms are coming from will help us predict where the most moisture will fall and collect.

Climate Analogs

With today's technology we can find similar climates all over the earth that we can study and see what has happened naturally in their part of the world. The plants and animals in that Climate Analog are likely to do well in our own area.

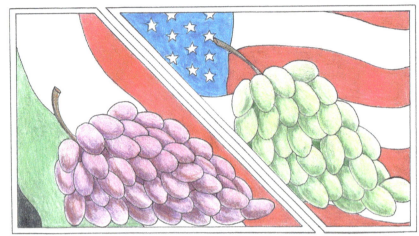

Grapes grow well in both Italy and California because they are both Mediterranean Climates.

Knowing what climate you are in allows you to easily compare yourself to other areas, but to find exact Climate Analogs takes some research. One way is to use the Koppen-Geiger Climate map online (found on Wikipedia.org).

Patterning

Natural systems consist of a series of overlapping **interrelated patterns**. We learn and communicate with patterns. Languages are patterns. Landscapes have repeating patterns as well. We learn with patterns, and we can learn nature's patterns as well with observation and study. The patterns for your area are outside, ready to be observed. People who have been in your area a long time know the longer cycle patterns and can tell you about them. These can be the largest rainstorms or floods, the driest, hottest summers, or the coldest winters. It all depends on your area and its conditions.

> **Interrelated:** to be connected to each other
> **Patterns:** a regular and recognizable process that repeats

Sun Path and Orientation

The path of the sun during the day changes throughout the year as the earth's tilt changes. The sun is the primary source of all energy on earth. If homes or gardens are inappropriately placed, they can get too hot or not enough sun. This does not make for a comfortable house or a productive garden.

Knowing the extremes and median of the Sun Path is vital to design.

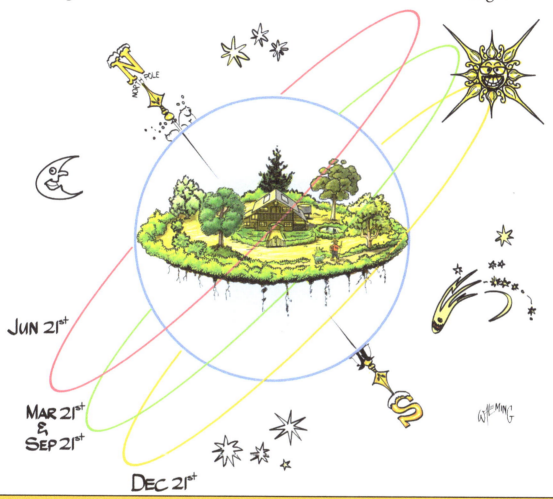

Solstice is the extreme of the sunpath - in summer the sunniest day and in winter the darkest day.

June 20th, 21st or 22nd - Summer Solstice in the Northern Hemisphere. Winter Solstice in the Southern Hemisphere.

December 20th, 21st or 22nd - Summer Solstice in the Southern Hemisphere. Winter Solstice in the Northern Hemisphere.

On or around March 21st and September 21st - The median of the solar path occurs twice a year.

PERMACULTURE DESIGN | 41

Slope

The steepness of an area determines what you can plant there. If it is too steep, the only thing you can do is plant trees and specific plants to prevent erosion. Annual gardens tend to be on the flattest possible ground available. Flat ground absorbs and holds water better than steep hillside likely to erode. The more flat or concave, the more water can be absorbed.

How to Calculate Slope

By imitating the picture or using posts with string and a level, we can measure the rise and the run to calculate the slope.

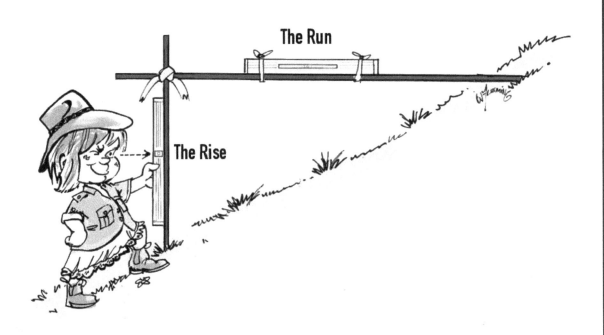

$$\left(\frac{\text{Rise}}{\text{Run}}\right) \times 100 = \text{Slope \%}$$

Edge Effect

Edge effect occurs when two different mediums meet. The species of both mediums combine with the edge species making a total of three times the amount of biodiversity.

For example the shore has ocean life, land life, and coastal life all in one place making it more fertile than the larger ocean or inland area. Edge effects can be purposefully created with swales, hedges, fences, & numerous other methods. The precolonial oak savannas of North America were a managed edge ecosystem of both forest and grassland.

Contour Lines

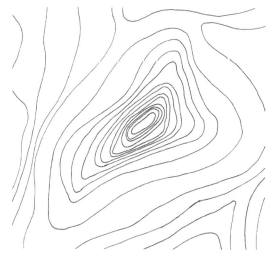

A contour line is a line of continuous elevation, so that any path along that line is perfectly flat. This is very useful in design. Water on a flat surface slows down until still and can soak in if on a **porous** surface. Contour lines have nearly endless application in design. Identifying the longest, highest, and lowest contour lines is especially useful.

> **Porous:** allowing air and water to pass through

Yield

Yield is the amount of product a system cre-ates. In permaculture there isn't a focus on a single crop yield per area. Instead, all the yields per area are counted because there are **polycultures** occupying the same space which is called **stacking**. The end result is more yield than any individual plant or animal could create. A good example would be the Three Sisters, a Native American **planting guild**, in which corn, squash, and beans are all planted together in the same area.

Nature has a dispersal of yield over time to support an array of niches and cycles. Year-round food is only possible if plants are diverse and mostly **perennial**.

Annual gardens are in addition to a perennial foundation. Having Early-, Mid-, and Late-Season varieties for each species spreads out harvest and yield, allowing for easier harvesting, longer.

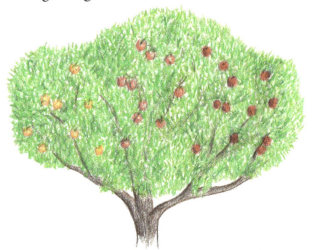

*his apple tree has early, mid, and late season varieties **grafted** to it.*

> **Polyculture:** a mix of several crops or animals in the same area
> **Stacking:** having multiple elements occupying the same space and/or time
> **Planting Guild:** a beneficial grouping of plants
> **Perennial:** a plant that lives many seasons
> **Annual:** a plant grown from seed each year
> **Grafted:** when a section of one plant is attached to another plant

Diversity, Stability & Sustainability

Dispersal of yield is an extension of energy over time by design. It increases diversity and stability in a system. It creates sustainability and as it gathers strength it becomes resilient.

Soil pH

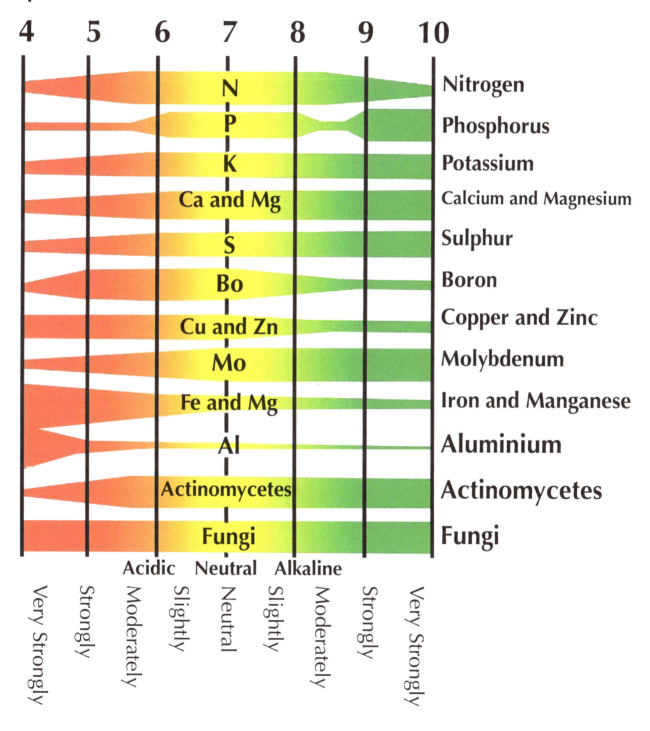

pH is a measurement ranging from extremely acidic (pH 1) to extremely alkaline (pH 14, though our chart only needs to go to 10). It measures the Hydrogen (H) ion concentration. Each degree higher in the scale is 10 times greater in concentration than the one before; it is a logarithmic scale. 7 is neutral, neither acidic nor alkaline - water is neutral.

Most gardeners strive for a pH of 6.5-7, but some plants do prefer a slightly more alkaline or acidic pH. Plants prefer the soil types they originated with in the wild. Testing the soil pH is somewhat misleading - pH changes every micrometer and can range widely. Compost tea and compost can help without even testing. If the soil is too acidic, the compost or compost tea will make it less acidic. If it's too alkaline, the compost or compost tea will make it less alkaline. Compost is the most effective soil amendment because it inoculates the soil with life.

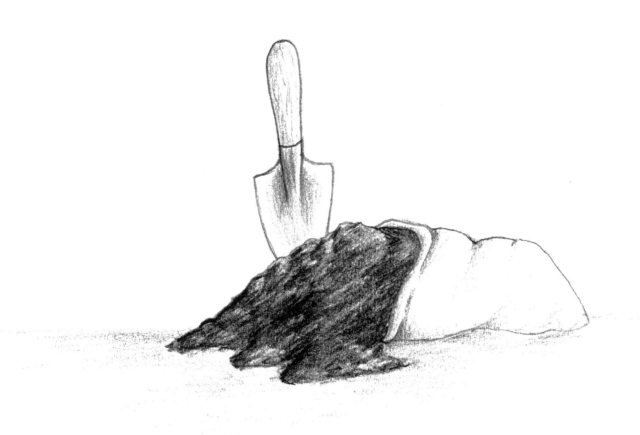

Planning

Functional Design

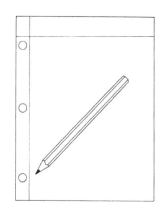

Functional Design is sustainable design that creates surplus yield. It connects as many elements in a system as possible to trap as much energy as possible on a site. Dysfunctional designs are not sustainable, require costly inputs, and break down eventually.

As a designer, we must always strive to connect every output to an input, to cycle nutrients and energy as many times as possible through a site, and include as many life elements as possible to build diversity, create stability,, and become regenerative.

Reading the Landscape

Every landscape has a story to tell. From the way trees lean caused prevailing winds to the floodwater lines around a seasonal creek, they all tell the history of that place. Though it takes time and practice, anyone can read the landscape. Using tools like maps, observation on-site, local research, and historical records, we can be ready to see what is waiting to be seen.

Topographic Maps

Topographic maps are made of contour lines that represent the physical landscape. They are not perfectly accurate. Seeing a site in person is the only way to know what will work there, but topographic maps make many things easier.

Topographic maps help find possible dam and house sites and clearly show which areas are too steep for anything but erosion control.

KeyPoint

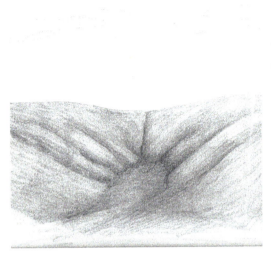

The Keypoint is the point just after which the land switches from convex to concave coming down from a ridge or mountain peak. This causes the silt, clays, and organic matter that was being carried by the water flow downhill to deposit as the water slows. This means more nutrients, more clay particles, and more natural water retention. It is the ideal location for water catchment or a dam.

Keyline

A Keyline is the contour line that extends away in both directions from the Keypoint. It catches the most water and has the most potential for regenerative designs. These keylines can be swales that both absorb water and divert it to the keypoint in flooding, or they can be non-absorbent and only divert water. It depends on your situation; in the desert it takes large amounts of catchment to irrigate a smaller area while the humid tropics would not have the need to do so.

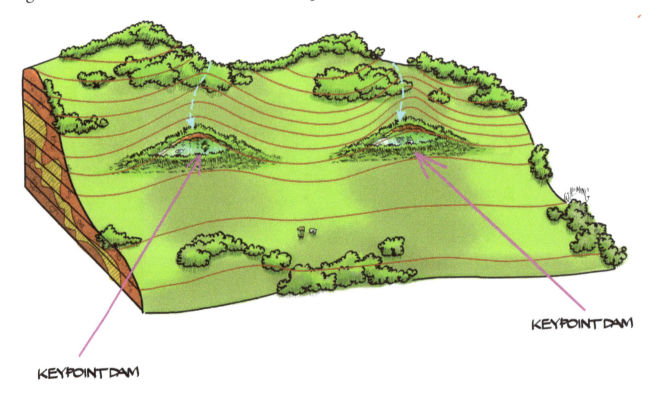

KEYPOINT DAM

KEYPOINT DAM

Calculating Catchment

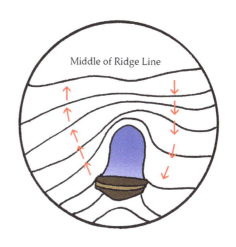

Using a contour map and starting from the pond site, trace at a right angle (90°) to contour until the ridge is met on both sides. The outlined area is the water catchment. Your local county records or town or city library will have the maximum rainfall historically recorded. It is the total area times the maximum amount of possible rain that calculates the maximum flow of water. Knowing this information determines the size of the pond, the dam wall, and its level sill spillway. Spillways are not always necessary in swale construction but are usually wise additions. They are lower than the dam wall, so water never rises above a certain level. This protects the dam wall.

> Area of Catchment **x** Maximum Rainfall in 24 hrs
> = Maximum Flow of Water

Analysis of Elements

Every element has needs, products, behaviors, and intrinsic characteristics. Mapping these out allows for a designer to see the full possibilities, strengths, and weaknesses of every element they consider. It is how each animal and plant are selected for every system initially. Experimentation is of course welcome, but good planning guarantees a yield and return on investment.

Intrinsic Characteristics
Breed, Coloring, Breed Specific Behavior, Climate Tolerance

Needs

Shelter	Grit
Water	Fresh Air
Food	Other Chickens

Products and Behaviors

Eggs	Flying
Meat	Fighting
Manure	Methane
Scratching	Feathers
Shredding	Foraging
Carbon Dioxide	Mulch Production

PERMACULTURE DESIGN

Sector Planning

Sector Planning is a method of minimizing the amount of energy spent maintaining the site. By organizing the different types of elements into zones, designers can put more high maintenance elements closer to the home to shorten the amount of steps per year taken to that element. Things like a kitchen garden are close to the home (Zone 1) while things that are rarely attended to like a timber forest are far from the home (Zone 4).

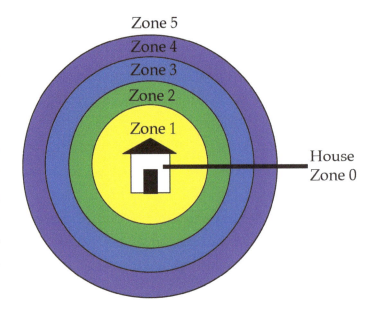

Zone 1: area around the home, herb and vegetable mulch gardens, most **maintenance** required
Zone 2: main crop, orchard, routine maintenance, small animals, animal **forage**, **dense** planting, heavily mulched
Zone 3: hardy trees, native species, animal forage, **grazing** and **browsing** animals, connects easily to zone 1 and 2, windbreaks, **firebreaks**, **rough mulched**, food forests, regular but not as intensive maintenance focused on animals, harvesting and cutting mulch.
Zone 4: timber, firewood, food forest, forage forest, minimal maintenance
Zone 5: wilderness, no maintenance, hunting, regrowth, timber

Maintenance: work to keep a system functioning
Forage: food that animals can self-harvest
Dense: close together, thick
Grazing: to eat grasses and pasture plants
Browsing: to eat leaves, twigs, bark, and other vegetation up off the ground
Firebreak: a fire obstacle like a open section of forest
Rough Mulched: large sections of foliage are cut and dropped to the ground without shredding or processing further

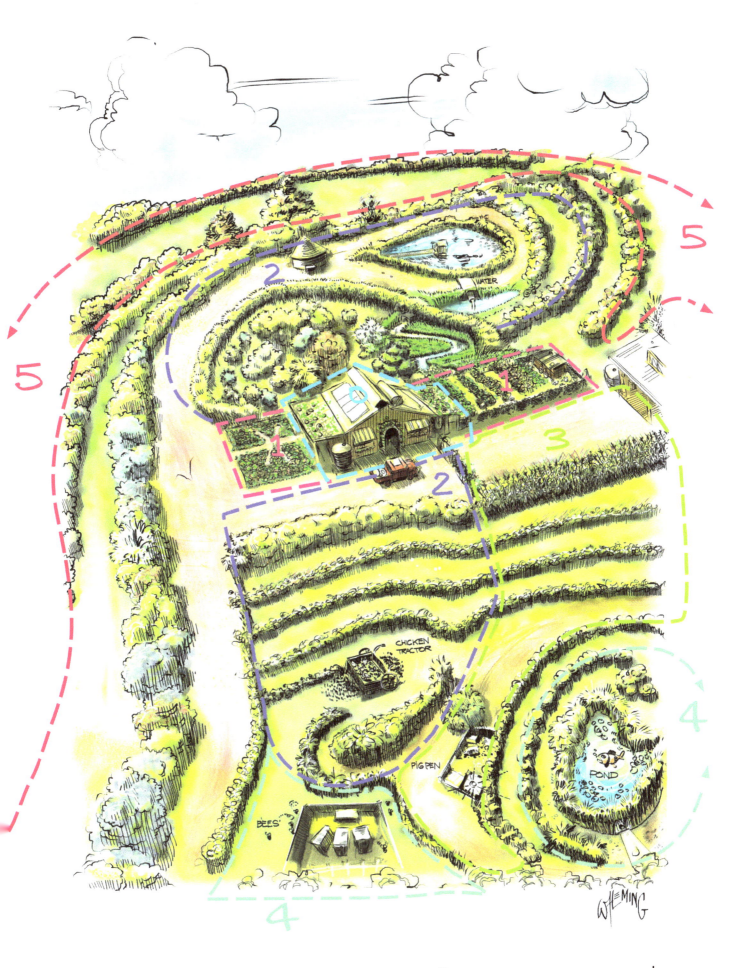

Random Assemblies

Random assembles are a way to generate ideas. They consist of listing possible features and possible interactions, and then randomly connecting two elements with an interactions. Though random, it does imitate the way nature generates diversity and can create surprisingly creative innovations.

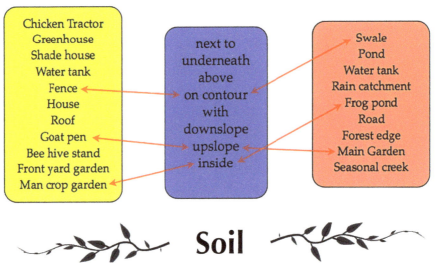

Soil

Jar Soil Test

The Jar Soil Test is a simple, easy way to discover the proportions of sand, clay, silt and organic matter in your soil. Combine half a jar full of soil with water, shake, and then let it settle. Measure the settled layers individually and as a group. Use the percentages to figure out your soil type with the soil chart.

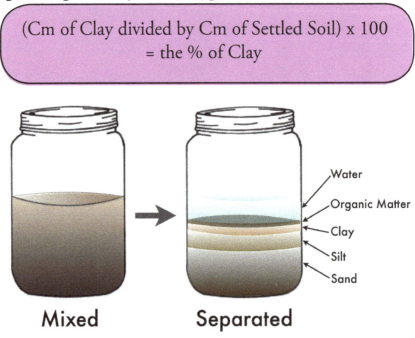

(Cm of Clay divided by Cm of Settled Soil) x 100
= the % of Clay

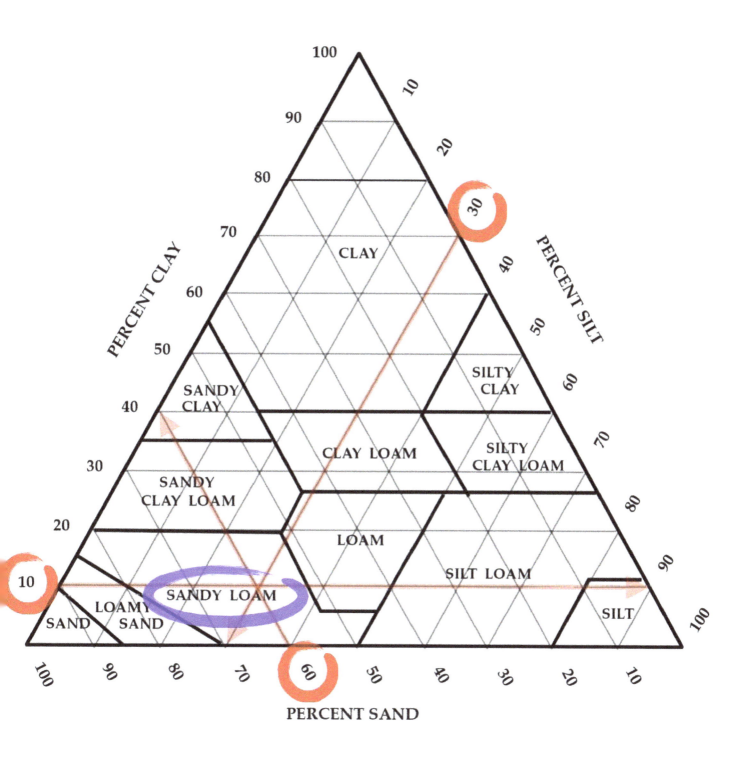

PERMACULTURE DESIGN | 53

Potting Soil for Starting Seedlings

It is important to remember that some seeds need **cold stratification** or **scarification** to **germinate**.

Potting
50% **sharp river sand**
50% **sieved** compost

Tropical
40% sharp river sand
60% sieved compost (or more compost)

Fine seeds
90% sand
10% sieved compost

Cold Stratification: a seed pretreatment that imitates winter conditions, keeping seeds cold and moist for a time

Scarification: when seed coatings are penetrated to allow water and air in. They can be cut or scratched. Hot water soaking and fire can also allow water and air in.

Germinate: to begin to grow

Sharp River Sand: Large particle sand found in the inside bends of rivers that crackles when squeezed in the hand and allows water to easily run through it

Sieved: run through a fine mesh, so only fine particles pass through

Rooting Cuttings

Cuttings need a moist, shady environment but not too wet because they will rot or the roots will not reach for the water and grow properly. Making a small greenhouse for just rooting cuttings is easy and inexpensive. Anything that allows light in and traps the moisture will work. Some people even use plastic bags over their planting pots. The potting soil should be 100% sharp river sand.

Softwoods and Semi-Hardwoods can take 3-4 weeks to root. Hardwoods can take several months to a year to take root. Transplant when the roots are healthy and 1-2 cm or 1 inch long. Water with only worm juice or compost tea.

Clay Seed Balls

Masanobu Fukuoka rediscovered this ancient technique for sowing seeds encased in clay and manure or compost as part of his Do-Nothing farming techniques. Mix the ingredients thoroughly and shape into balls. Allow to dry in the sun.

Recipe
1 part seeds
3 parts compost or manure
5 parts clay
1 to 2 parts water

Sheet Mulching

Sheet Mulching is a soil building technique where cardboard, newspapers, paper, manure, and mulch are layered to build soil quickly. It is an imitation of the forest floor. Nature creates a thick mulch layer in the forest. That thick layer has all the potential of a new forest or meadow in it at all times.

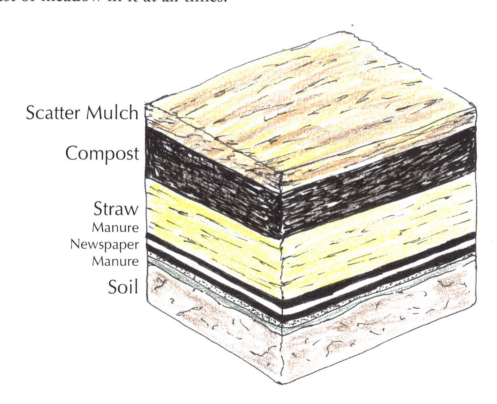

Sheet Mulching creates a consistent fungal layer with the decaying, wet wood fiber in the cardboard, paper, or newspaper layer. The animal manure provides bacteria, water holding capacity and Nitrogen (N). The straw/mulch layer creates cooler soil, holds moisture and air, and adds Carbon (C) to the soil. All these ingredients add more than what is listed, but these are the main active parts in the process.

Rough up the soil you are going to be sheet mulching with a rake, hoe, or anything that disturbs the soil enough to open it to moisture and then add in layers from bottom to top:

- Soil Amendments if necessary
- 2.5 cm (1") manure
- 1-2 cm (.25-.5") newspaper or cardboard
- 2.5 - 5 cm (1-2") manure, preferably with no seeds
- 15 - 25 cm (6-10") of organic mulch like straw, other dried carbon-rich remains of plants and even non-allelopathic tree mulch
- 2.5 - 5 cm (1-2") compost
- Scatter seedless mulch lightly atop to shade growing seeds, hold them in place while watering and hide them from predators.

Over time more mulch needs to be added, but if plants that are in position can provide the mulch, it is less work and a better design. Small leguminous shrubs that regrow quickly are perfect for this as are mineral accumulators. Comfrey, an herbaceous perennial, is a nutrient accumulator with deep roots. Planting it around fruit trees is one way to make easy mulch and healthier fruit.

Compost

Compost is a dark, rich, sticky, blackish-brown, soil-binding organic matter composed of long Carbon (C) molecule chains that bind a diversity of elements together in their chains. It is life-rich organic material broken down to humus. Composting is the action of breaking down organic materials into long carbon chains in a process of **decomposition**.

Compost is extremely useful. You can place it in the garden beds in pockets for seedlings to be planted, around established plants, as the top layer of a garden bed, and in compost tea. Its long carbon chains hold a great selection of minerals and nutrients that feeds into the food soil web. Healthy plants make healthier food for people and animals.

"If it has lived, it can live again... in the compost"
-Geoff Lawton

Hot Compost

Every hot compost has 2 basic elements that create the reaction: Carbon (C) and Nitrogen (N). High carbon materials, often called browns, are things like straw, wood chips, paper, or leaves. Animal manure supplies the Nitrogen needed for the reaction. A 25 parts carbon to 1 part nitrogen ratio (25:1, C:N) is needed for a hot compost reaction to reach the right temperature. The heat indicates that microbes are hard at work breaking down the individual materials into **uniform** compost. A compost heap's ideal temperature is between 131-140°F (55-60°C) for at least 15 days. This temperature level kills harmful **microbes**, pathogens, and weed seeds. When it gets too hot, it's time to turn it, release the heat and start over. Turning regularly helps aerate the pile which fuels the reactions inside. If a pile goes anaerobic, it is not getting enough air and will smell bad. Aerobic reactions smell earthy but not putrid.

> **Decomposition:** the process of rotting, decaying
> **Microbes:** small living things seen only with a microscope
> **Uniform:** appearing the same

18-Day Berkeley Compost

The 18-day Berkeley compost method, developed by the University of California Berkeley, is a fast and reliable way to create high quality garden compost. The minimum size for the pile is a cubic meter (or almost 3.5 cubic feet). It needs to be at least that size to get hot enough. In addition to browns which are plants that have gone to seed and manure, there are greens which are plants that have not gone to seed like fresh cut grasses or weeds: these add more microbiological diversity to the process. Dead animals or fish, comfrey, nettles or old compost can be added to the center of the compost in the middle at the very start of the compost process. It will speed up the heating process and bring an increased micro-biodiversity to the finished product.

Berkeley Compost
1/3 High Carbon (shredded)
1/3 Greens
1/3 Manure

18 Day Berkley Compost Schedlue

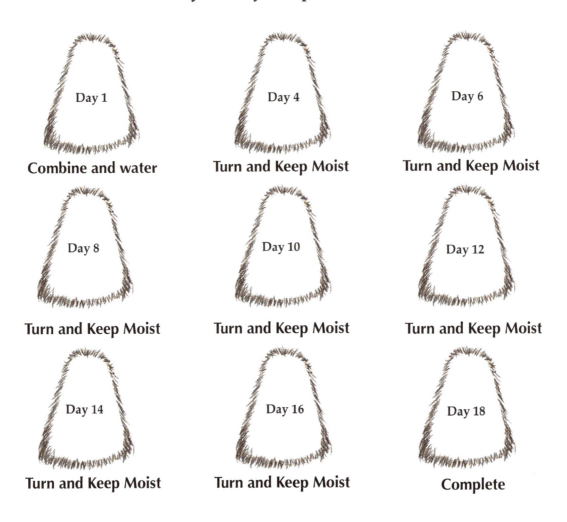

Day 1	Day 4	Day 6
Combine and water	Turn and Keep Moist	Turn and Keep Moist
Day 8	Day 10	Day 12
Turn and Keep Moist	Turn and Keep Moist	Turn and Keep Moist
Day 14	Day 16	Day 18
Turn and Keep Moist	Turn and Keep Moist	Complete

When you have your materials ready, start making your pile in layers with Browns first (to keep it well aerated). Once it is built, water it until it leaks and turn it on schedule (or when it gets too hot or too cold - at least 5 times in 15 days). Make sure it is thoroughly wet in all areas; watering while building can be an option. Check the moisture levels routinely to keep the reaction running well. If you squeeze a handful of the pile and it only drips a few drops, that's the right level of moisture.

Compost Tea

Compost Tea is a liquid full of soil life, not a plant fertilizer. It is created by putting compost in a mesh bag, suspended in a bucket of water, and then aerating the water for 12 - 72+ hours while the soil life separates from the soil particles and breeds. The end product is a life-rich aerobic liquid that brings back lifeless soil and helps plants thrive as a result of healthier soil life. There are many ways to make a Compost Tea Brewer and many recipes that depend on the needs of the soil and plants, but the basic ingredients are compost in a mesh bag, water in a bucket, and an aeration method (like an aquarium pump). People also add ingredients like molasses, kelp, trace minerals, fish hydrolysate, and other microbial foods to add more nutrients and minerals or to influence the fungal:bacterial ratio of their compost tea as well.

Once ready, use compost tea within 6-8 hours. Dilute the tea before using to 1 part tea to 2-4 parts water until it is a weak tea color. Apply to the soil once a growing season on average.

Worm Juice Compost

A worm juice compost is a system for creating compost without consistent maintenance. It is perfect for kitchen scraps which are perpetually created.

Any container can work as long as it has a way to drain out the bottom. The inside of the container is raised, so that the manure and compost do not touch the bottom of the container. Shade cloth over a frame and stilts or gravel can hold up the compost, allowing it to drain liquid.

Once gravel or shade cloth are in place, put down a thin layer of dry straw or leaves and then fill the container half full with manure and add worms. Fill the rest with kitchen scraps regularly. Worms will digest and convert the materials into worm castings. The worm casting juice has great bacteria for the soil that can be added continuously throughout the growing season, and in 3 months time, the entire container is ready as garden compost. If it ever stinks, hot compost it before putting it in the garden.

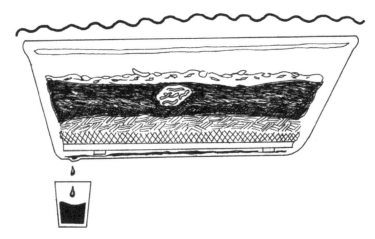

Biofertilizer

Biofertilizer is a non-living fertilizer for plants and soil made in an anaerobic fermentation process. This fertilizer is especially good at adding minerals back into the soil and plants. There are many ways to make a biofertilizer system.

Combine all ingredients in a 50 gallon barrel. This can be a used feed barrel. The top and air tube connections must be airtight. The air tube feeds out into a water bottle or any container filled with water, so the air tube is always under water. No need to seal this part. As it ferments, the water inside the bottle will bubble with released gases from the fermentation process inside the barrel. In 3 months, you have golden liquid fertilizer that can store indefinitely but will require aeration like compost tea to remove any potential remaining anaerobes. Water down 20 parts water to 1 parts biofertilizer when applying to plants and soil.

Biofertilizer

10 kg/22 lbs rumen (1st and 2nd Stomach) or fresh manure
10 L/2.5 gal molasses
2 L/.5 gal milk
5 L/1.3 gal kelp
1 kg/2.2 lbs brewer's or bread yeast
1 kg/2.2 lbs powdered double burnt bone

Plants

Chop and Drop

Chop and Drop is as simple as it sounds but has huge benefits. When people pull up weeds and remove them, they are removing the nutrients that soil needs that the weeds are accumulating. When we chop them down, chop them up, and leave them, we speed up the natural repair process. The smaller the pieces, the more surface area, so the faster it breaks down.

Legumes

Legumes with soil bacteria accumulate nitrogen in the soil. Legumes enrich the soil, so that other plants can thrive. They also come in a range of varieties, covering all the layers of the forest. They are fast-growing, often called "weeds". Legume trees can be **coppiced** or **pollarded** without killing the plant.

PERMACULTURE DESIGN | 61

Legumes can be used in many ways. They can prepare the soil for a garden or a food forest as a **cover crop**. They can be **support species** in a food forest. They can be food for people or animals. Their wood can be used for firewood as well. They can also be used as superb mulch.

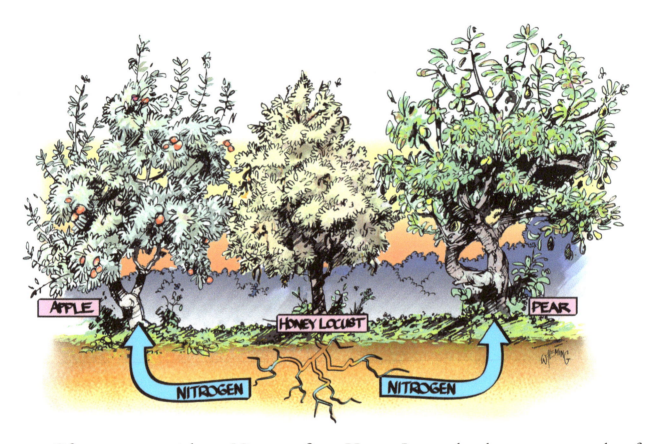

Often controversial as a Nitrogen fixer, Honey Locust has been proven to be effective in commercial settings as on Les Fermes Miracle Farms in Quebec, Canada, and the debate has opened a branch of study into primitive N-fixers that lack the telltale symbiotic N-fixing bacterial root nodules of most N-fixers like Black Locust, Siberian Pea Shrub, or Acacia.

> **Coppiced:** cutting trees at their base to promote vigorous regrowth. Some trees such as conifers do not coppice well.
> **Pollarded:** cutting a plant's top or top branches to encourage growth
> **Cover Crop:** a plant grown to cover bare soil that often enriches the soil, commonly legume annuals used to rest land between intensive annual spring and summer plantings.
> **Support Species:** an animal or plant that supports the existence of other plants and animals

Planting Guilds

A planting guild is a group of plants, a polyculture, that work together well. They improve and protect each other's functions. We can research guilds, companion planting lists, and local gardening successes to learn more.

Food Forest

A food forest is a designed tree landscape that is built by imitating the natural processes of forest development. Using legumes, chop and drop, planting guilds, desirable forest layers, and earthworks, a food forest can be established quickly and last hundreds if not thousands of years.

Generally, the planting starts out as 90% support species and 10% productive trees, and at climax, it is 10% support species and 90% productive trees.

Fast Ecological (or Forest) Succession

Legume Cover Crop - 6 months
Small Legume & Valuable Bushes - 4-5 years
Medium Term Legume & Valuable Trees and Bushes - 10-15 years
Full Term Legume & Valuable Trees - 15-30 years

Legumes provide a faster succession rate while feeding plants & the soil. This leads to less work & higher yields early.

Net and Pan

Net and Pan is a tree planting system used in dry climates and steep slopes. Trees are planted in shallow **depressions** (the pan) in the soil and connected by a network trenches (the net). This catches runoff and rain water and delivers it to the tree guilds along with any nutrients or mulch gathered along the way.

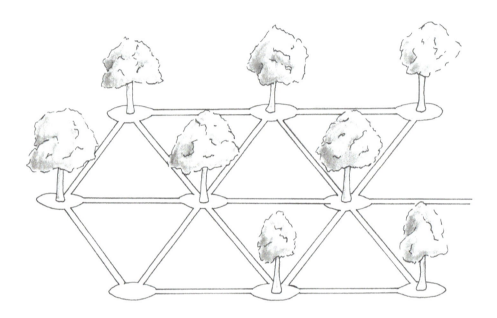

Mass Planting

Mass planting selection is a plant breeding technique. When planting any plant in a very large grouping, it reveals all the **genetic variation** possible in that plant. This makes it easy to find desirable **traits**, even if they are only found in 1 out of a 1000 plants. The probability of finding rare genetic traits goes up as the size of the plant population goes up.

Selecting for desirable traits should be done carefully. Breeders should select for no more than 2 traits at a time. Early yield and **vigor** should be the first two traits selected for. Later, once those genetics are established, other characteristics like taste, color, and high yield can be selected for.

> **Depressions:** lower areas in the land
> **Genetic Variation:** the diversity of genetic expression
> **Traits:** a characteristic
> **Vigor:** strength and abundant health

Windbreak

Windbreaks provide shelter from wind and can be made of almost anything. Rows of hardy trees make effective and sustainable windbreaks. Fences, hedges, or berms can also serve this purpose. On a site with strong winds, windbreak is very important.

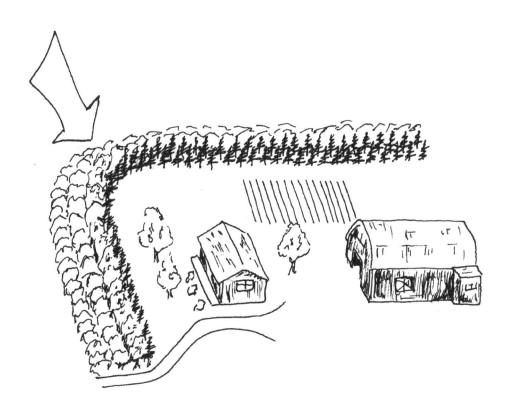

Microclimates

Microclimates can be designed to stretch the growing possibilities on a site. The options for creating a microclimate are limitless. Rocks and ponds absorb the sun's energy and radiate that heat long after the sun has set. They also reflect the sun's light onto other objects. Windbreaks prevent wind from cooling or heating a site. Orienting towards the sun provides the most potential energy for that site, but in extremely hot areas, shade may be what's required. Microclimates are a manipulation of the amount of energy coming into an area.

Animals

Paddock Shifting

Paddock Shifting is an animal grazing technique that improves soil, pasture, and animal health. Animals are at a grazing location a short time before they are moved to a new location, usually just a day. This lessens compaction, improves pasture regrowth, and prevents the animals from eating too much of the pasture or plants that are not as healthy or bad for them. Because the animals only eat the best of the pasture, their health and nutrition improves steadily over time.

Chicken Tractor

Chicken Tractors are portable habitats for chickens that allow for foraging. It is similar to paddock shifting, but instead of the animals moving alone, the structure moves with them.

Many animals can be "tractored", like cows, pigs, sheep, goats, and rabbits, though the smaller animals have smaller structures, so they are easier to move. More often pad-dock shifting is used for large animals and tractoring for small animals but not always - according to some historical records, Thomas Jefferson used a cow tractor on his farm in Virginia. Animal tractors work best on flatter and more uniform ground but adaptations can be made to suit almost any landscape profile.

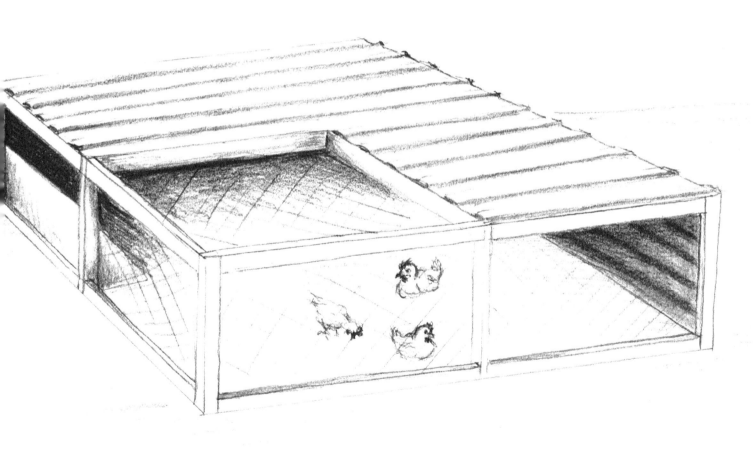

Aquaculture

Aquaculture is the raising of **aquatic** animals and **cultivation** of aquatic plants for food. These systems can be many times more productive than land-based systems. The larger the body of water, the more stable the life systems can be which means less work maintaining it but more work harvesting. Aquatic plants make excellent mulch. They hold larger amounts of water than land-based plants, in some cases holding up to 40 times their weight in water.

> **Aquatic:** belonging to a water habitat
> **Cultivation:** the process of growing

The Aquaculture Chain of Life
- Algae
- Zoo Plankton
- Crustaceans
- Fish

Water Plant Levels
- Edge Plants
- Shallow Water Plants
- Deep Water Plants
- Floating Plants

Chinampas

Chinampas are the most fertile and productive food systems in the world. By combining aquaculture and perennial agriculture in a design that enhances edge effect, these systems can stay fertile for centuries. When the Spanish arrived in the Valley of Mexico for the first time, they witnessed an amazing network of aquaculture water canals with crops growing on the island strips between them. The land strips were held together with fences and trees like willow and cypress.

A chinampa is created by digging into the soil below the shallow water and creating a trench while piling that soil in a mound next to it, deepening the water and raising the land. Once the land is higher than the water it can begin to dry out. The airless soil below the water is anaerobic and needs time for the soil life to become aerobic. Shallow marine waterways have very fertile soil, so growing food in a chinampa is highly productive. It is also a prime example of edge effect. Since it is essentially all edge, the species interaction increases dramatically. The soil is enriched continuously which leads to higher production as well.

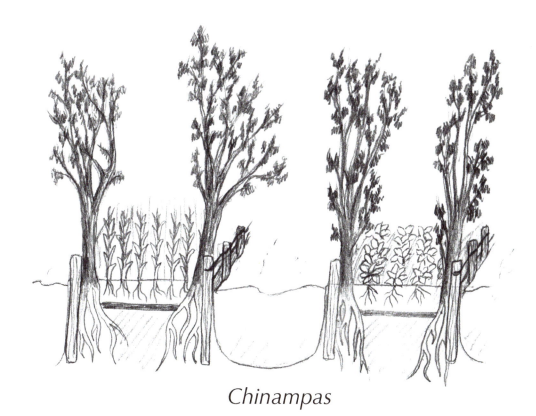

Chinampas

Ponds

Ponds hold water and create diverse life systems that can raise the fertility and yield of a site very quickly. If pond systems are connected to land systems, even more possibilities open up.

Fish that can eat a vegetarian diet like Tilapia can be fed the plants in and around pond. They self-harvest the foods, and their wastes feed the plants in a continuous cycle.

> **Hydroponic:** plants grown in nutrient-rich water without soil

Aquaponics

Aquaponics is a fish and **hydroponic** food system where fish waste is filtered out by plants. It is a simple cycle of pumping waste water into gravel beds with plants in them and then running that water back into the fish tank. There are many variations of this basic model. If you grow plants that the fish eat, the plants and fish will feed each other indefinitely (as long as your equipment works!)

Clean Water Pumped Back to Tank

Earthworks

Earthworks are land manipulations by design to capture more direct and potential energy.

Hugelkultur

Hugelkultur, or *mound cultivation*, is a raised garden bed technique that imitates forest cycles. Just like how a forest turns into soil, the hugelkultur decomposes dead trees buried under mulch and soil. Moisture is held by the rotting logs inside, heat is generated as it breaks down, and carbon and nitrogen are steadily released into the soil and taken up by the plants growing on the hugelkultur.

Hugelkulturs tend to have a more shaded side and a more sunny side. This helps make planting choices easy as many plants have a preference between full sun and partial shade.

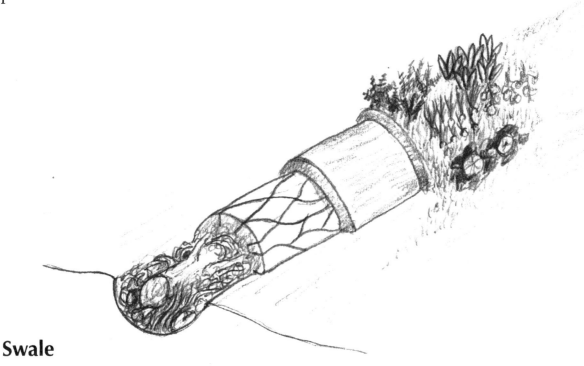

Swale

A Swale is a ditch that is on contour, so that it absorbs water passively into the landscape. It can be made by hand, with a shovel, or with a large excavator. The scale does not change the function of water absorption. Using a tool to find level and stakes to mark the contour line, the ground can be removed from the uphill side and placed on the downhill side of that line. The back cut of the swale path should be at the same angle as the slope after the swale mound.

Swales stop water and force it to soak into the land because they are flat surfaces with uncompacted mounds of soil below them. In heavy rains they can fill up, so it is important to include **level sill spillways** for safe, even overflow. This protects the swale mound from erosion and possible mud slides.

> **Level Sill Spillway:** a section of compacted dirt that is lower than the dam or swale wall that allows overflow before water gets too high in the dam or swale. It allows water to gently flow over it in a thin, even sheet to prevent erosion.

Swales are tree planting systems and should be planted immediately or soon after excavation to prevent erosion. The majority of plants will be legumes, but any nitrogen fixing plants that work well in your area and in your design can be used. Hidden among the nitrogen fixing plants will be valuable fruit, nut, and timber trees. The legume trees will be cut back routinely to feed the valuable trees with their mulch on the surface and proportionately dying roots which both fix nitrogen and carbon below the surface. The smaller elements will die off or grow along the edges, leaving longterm large legume trees as companions to the longterm, valuable fruit, nut, and timber canopy. These trees will hold the swale in place for generations and create longterm shade and windbreak, retaining moisture and warmth in the ground longer, which will build fertility and diversity. Please note, swales can work too well: they can make an area overly damp. Avoid using them in damp, heavy, clay soils - subsoil rip instead.

Dams

Water is a precious resource. Only 3% of the world's water is fresh water, or water that is not salty. 75% of that fresh water is frozen. The remaining available water needs to be managed properly with a well-planned design. Permaculture provides the way that water can be retained in the land for both our use and the land's.

"Where there's water, there's life."
- Geoff Lawton

Dams or ponds hold water in the landscape. The most common dams are found in valley bottoms. These have the most catchment but also the most pressure on their walls, and they lack potential energy. Using gravity can be very powerful. A pond can still occur in the lower landscapes, but in a good design that is only after that flow of water has expended all its potential energy coming down the slope.

Dam wall width to the full dam length should be a ratio of 1:3, wall width:dam length. That's why designer's look for pinch points in the landscape to save money, time, and energy.

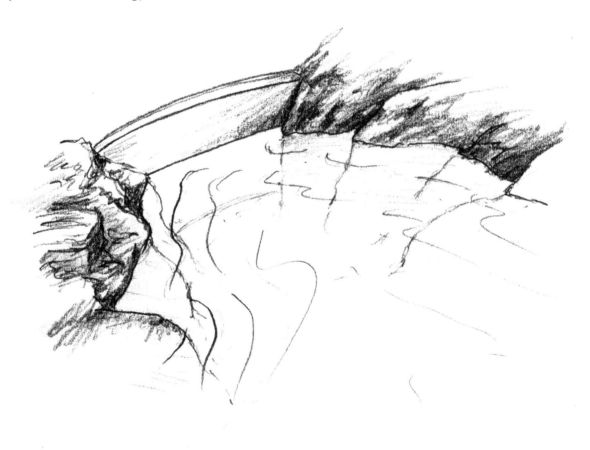

Earth Tanks
- Flat lands
- Water storage
- Water has to be pumped in

Contour Dam
- Built on low, flat land, <8% slope
- On contour
- Flat bottoms
- Shallow
- Aquaculture

Keypoint Dam
- Reforestation technique, connects valley catchments
- Built at the keypoint where the slope changes from convex to concave.
- Often connected by swales along the Keyline
- The Keyline Swales often connect to other valley Keypoints

Ridge Point Dam
- Flat part of ridge
- Can be connected to swales
- Higher wall on ridge point

Saddle Dam
- On a ridge between two hills
- The highest dam
- Two walls
- Spillways anywhere
- Can be connected to swales

Gabions

Gabions are wire containers, often cubes, filled with rocks or broken concrete for dams, erosion control, or other construction purposes. They trap silt behind them while the rocks condense water. This condensed water can create a steady, usually small, stream of water for periods of time. In very dry areas, a series of gabions down a slope might be the only source of water for miles in all directions. The top gabion could have a small trickle for three months, the next gabion for six months, the next gabion for nine months, and the final gabion year round. Depending on the site it may take more or less gabions to collect enough water in the described example.

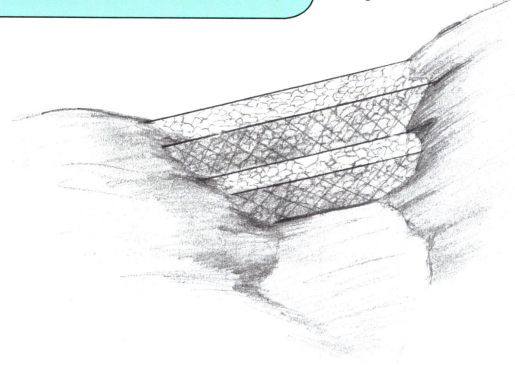

For the Home

Rainwater Catchment

Any hard surface like a roof is going to have complete runoff of rainwater. Using gutters, a **first flush rainwater diverter system,** and a water tank, nearly all the rainwater can be captured and stored.

> **First Flush Rainwater Diverter System:** The first rains wash the roof. This system diverts that wash water away every time it rains, so that the water collected in the main tank is relatively clean.

Rocket Mass Heater

A Rocket Mass Heater is a rocket stove running its **exhaust** through a thermal mass like a stone, concrete, sand, or **cob**. It could heat up a bench, a floor, or a center wall. A rocket stove is a J-tube for stick fires that burns cleanly. The bottom of the J-tube is upright which allows for gravity to pull the sticks into the fire. The tall chimney pulls the air into the J-tube, the flame burns sideways, and it creates a rocket re-burn effect where the exhaust becomes the fuel in the tall upright part of the J-tube. The clean heat that comes out is channeled into a mass to store the heat and let it radiate slowly over time. Rocket Mass Heaters can be run for short periods of time and heat homes for over a day in winter in places like Montana, USA. Rocket stoves can also be used to create hot water or steam and to cook food with.

> **Exhaust:** gases released in combustion or in the operation of any machine.
> **Cob:** a natural building material that can be composed of water, sand, clay, and straw. It is fireproof and easy to shape into any form.

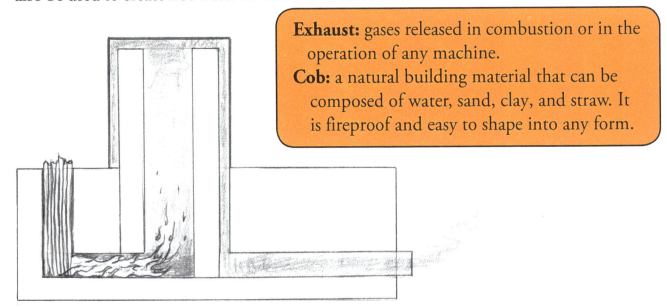

Greenhouse

A greenhouse is a structure with glass or plastic walls and roof designed to allow as much light in and trap as much heat as possible. Sometimes greenhouses can get too hot and need to be vented.

A well designed greenhouse can grow food all year round. It can also grow things that don't grow normally in your climate.

In addition to food, a greenhouse attached to the front of a home can heat it by venting the hot air into the house as long as the house faces the sun. A vent is placed high in the wall they share since hot air rises and will flow into the vent passively. A moisture barrier is vital to prevent mold and mildew.

Shadehouse

A shadehouse is a shaded structure for growing heat and light sensitive plants in hot times or climates. It also can be used to cool a house by installing a low vent on their shared wall since cooler air falls.

A shadehouse is attached to a home on the side that does not face the sun and is already shady. It requires a moisture barrier as well.

Walipini

A walipini is an underground greenhouse. In Aymara, a Bolivian indigenous tribe, 'walipini' means 'place of warmth'. This design uses the thermal constant of the earth and sunpath orientation to keep plants warm in extremely cold climates. Its roof is clear plastic or glass. The roof angle is 90° to the angle of the sun on winter solstice to capture the most energy on the darkest day of the year.

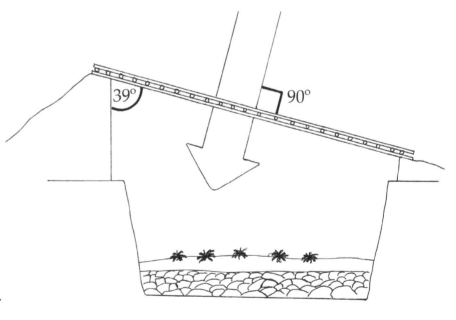

Walipinis have been used to grow bananas in winter in the Andes at 6,000 ft elevation. They can trap a lot of heat. Many walipinis have chimneys to vent excess heat. Growing beds are on top of gravel to prevent water from becoming stagnant. They are an easy and inexpensive way to grow food in cold winters.

Wofati

Based 80% on Mike Oehler's work on earth sheltered homes, Paul Wheaton's wofati design is an earth-sheltered building that allows in plenty of light with the benefit of not needing air-conditioning or heating. This design traps the annual thermal heat of the summer sun and extends it into winter. The earth around them keeps them cool in summer and warm in winter. An added benefit is that they are quick and inexpensive to build.

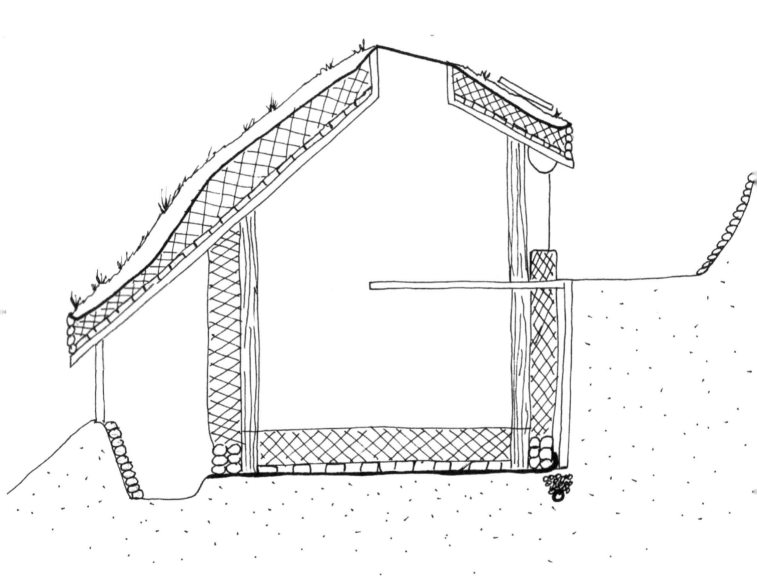

Chapter IV

Permaculture and the Future

 # Permaculture and the Future

If we can create a symbiotic relationship with nature, we can build the resiliency we need for a bright future. Using permaculture, we can reverse soil degradation, water scarcity, deforestation, pollution, hunger, and, in tandem, resource conflicts. We can go further and build resilient systems that will weather and protect us through climate change. It will take a worldwide effort with us each in our own communities doing what we can with what we have. All waste must be recycled on our own home sites. Energy and food need to be generated regeneratively and locally. We don't need to export or import anything. We just have to look around us; all our problems can be solutions. With the knowledge in this book, you can start to regenerate degraded and broken ecosystems. You can create abundance anywhere - no matter your age or circumstance.

Observe what plants you already have in your area. Do you have pioneer species of legumes? Can you collect their seeds? Can you harvest water from your roof or the land? Can you dig a swale? Can you collect mulch, rainwater, or organic matter? If you can do these things, you can start a permanent system that can start the healing process in your area.

Just start now!

MP

Index

A

Altitude Effect, 38
Aquaculture, 22, 33, 37, 70

B

Bacteria, 16, 24-25, 27, 56, 60, 62

C

Chicken Tractor, 69
Chinampa, 70-71
Chop and Drop, 31, 61, 63
Climate Analog, 40
Compost, 14, 18, 54-60, 69
Continental Effect, 39
Contour, 43, 49, 72, 75
Cycle, 14, 17-21, 25, 27, 31, 40, 44, 47, 71-72

D

Dams, 74-75
Diversity, 6, 16-17, 23, 29, 33, 43, 45, 47, 52, 57-58, 66, 73

E

Edge Effect, 42, 70
Energy, 6, 8, 10-11, 19, 21-22, 33, 37, 41, 45, 47-48, 50, 67, 72, 74, 77, 80
Ethical, 6-8

F

Fertility, 16, 19, 24-25, 33, 71, 73
Fish, 58, 70-71

Food Forest, 10, 50, 62-63
Fungi, 10, 12, 14, 24-25, 27-28, 45

G

Gabion, 75
Garden, 10-11, 17, 41-42, 44, 46; 50 57-58, 60, 62-63, 69, 72
GMO, 12
Gravity, 6, 33, 37, 74, 76
Greenhouse, 54, 77
Guild, 44, 63, 66

H

Herbicides, 10, 12, 24-25
Hugelkultur, 72

K

Keyline, 48, 74
Keypoint, 48, 75

L

Legume, 10, 16, 61-63, 73, 80

M

Maritime Effect, 38-39
Microclimate, 33-34, 67
Mineral, 10, 13, 24-25, 27, 33, 56-57 59-60, 69
Mulch, 18, 29, 31, 49-50, 55-56, 62 66, 70, 72-73, 80

N

Niche, 17, 30, 44
Nitrogen, 10, 16, 24, 45, 56-57, 61-62, 72-73

O

Organic, 12, 18, 23-25, 27, 52, 57, 69, 80

P

Paddock Shift, 68-69
Pattern, 6, 13, 17, 37, 40
Pesticides, 10, 12, 24-25
pH, 45-46, 69
Pollution, 14, 80

R

Rain, 21, 39-40, 49, 66, 73, 76, 80
Rain Shadow, 33, 39
Rocket Mass Heater, 76

S

Sector Planning, 50
Seeds, 10, 16, 18, 23, 54-57, 80
Shadehouse, 77
Slope, 33, 42, 66, 72, 74-75
Soil Food Web, 25
Solar, 8, 41
Sunpath, 41, 77
Swale, 49, 63, 72-73, 75, 80

W

Walipini, 77
Water, 6, 8, 13-14, 16, 19-24, 29, 33, 37-39, 42-43, 47-49, 52, 54-56, 58-60, 66, 69-77, 80
Weeds, 10, 31, 58, 61
Wind, 6, 8, 23-24, 29, 33-34, 38-39, 47, 50, 67, 73
Wofati, 78

Y

Yield, 44-45, 47, 49, 66, 71

About the Author

Matt Powers teaches permaculture and regenerative gardening and farming to families, youth, schools, and adults all over the world through his online courses, videos, and books, which are now being used on all continents (except for Antartica.) Translations of his books are already available in Spanish, Polish, and Arabic with French and Italian coming soon. As an experienced educator with a masters degree in Education, Matt went from teaching high school students to teaching high school teachers and administrators to teaching districts and appearing at universities and conferences all over America and online, teaching permaculture and sustainable, regenerative skills and thinking. Matt provides daily inspirational and regenerative content online and is one of the most-followed permaculture teachers online with over 30,000 Twitter followers and tens of thousands of followers in his many Facebook groups and pages ranging in topics from permaculture education to entrepreneurship to gardening to fungi & more.

Visit Matt Online:
ThePermacultureStudent.com
Twitter.com/Permaculture123
Facebook.com/ThePermacultureStudent

Other Books by Matt Powers
The Magic Beans
The Permaculture Student 2
Permaculture for School Gardens

"With The Permaculture Student 2 Matt Powers has completed a great service to the younger minds of the planet — others too, especially those new to land management, will find this piece very instructive, useful, easy to read and navigate."

— Darren J. Doherty, Regrarians.org

"Whether you are new to permaculture or well-steeped in its principles, The Permaculture Student 2 condenses a wealth of information into an easily digestible and thoroughly engaging format that is sure to give any reader new points of reference and insight. I highly recommend it."

— Peter McCoy, Radical Mycology

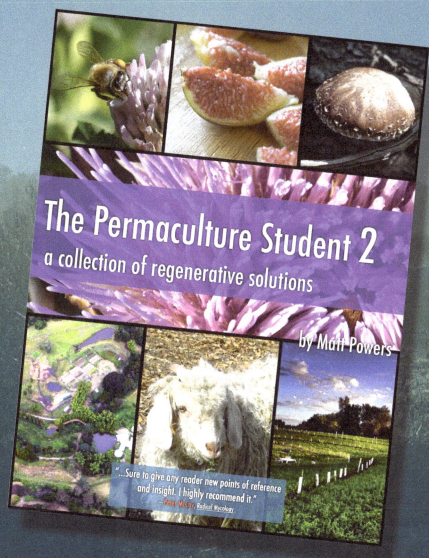

Order Yours Today from Amazon, Barnes&Noble, or Your Local Bookstore!